# THE GEOLOGIC STORY
# OF BERNHEIM FOREST

## By PRESTON McGRAIN

Kentucky Geological Survey
University of Kentucky, Lexington

# Table of Contents

## COVER PHOTOGRAPH

View looking west from fire tower in Bernheim Forest, showing portion of the forested hills and the belt of Knobs. The upland in the distance is the escarpment called "Muldraugh's Hill." This is the land from which the Knobs were carved by erosion. This and other photographs in this report courtesy of Bernheim Forest staff.

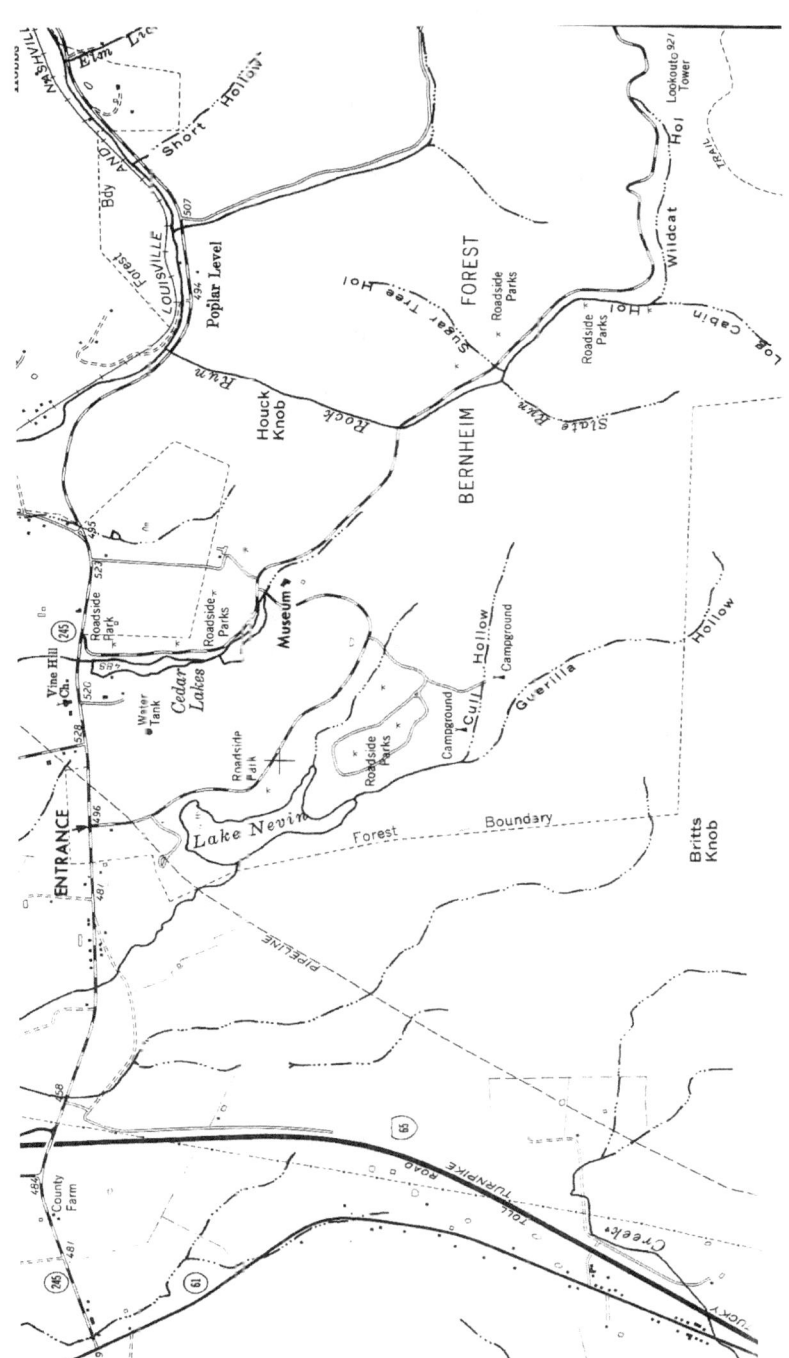

# Introduction

Bernheim Forest, a privately endowed 10,000 acre woodland, upland game sanctuary and natural science and horticulture educational area operated by the Isaac W. Bernheim Foundation, is located in Bullitt and Nelson Counties, Kentucky, about 25 miles south of Louisville, Kentucky. This is a region of scenic forested hills, verdant valleys, and picturesque knobs.

Although better known for the display of flowering trees and variety of vegetation, the Forest contains a variety of rock types and the student of natural science and the nature lover can find some interesting pages of the Earth's history revealed in the quiet glens and on the rugged hillsides. The combination of natural features has resulted in an interesting and instructive outdoor museum, an educational resource for Kentuckians and visitors from neighboring states.

Geologically, Bernheim Forest is in that part of Kentucky where the Blue Grass meets the Knobs (see Fig. 1). In north-central Kentucky, a horseshoe-shaped belt of ridges and hundreds of more or less isolated hills comprise a very prominent feature which borders the State's Blue Grass region on the west, south, and east. This upland area has been referred to as "The Knobs" or merely "Knobs" and constitutes a striking and beautiful portion of Kentucky's landscape.

This is also "Lincoln country." Abraham Lincoln's boyhood home was in the wooded Knobs region of nearby northern Larue County, in a setting very similar to that seen in the Bernheim area.

This booklet is designed to provide general geological background material for the visitor to the Forest. For those concerned with more technical aspects of the geologic features and history, several scientific references are included at the end of the discussion.

# Rocks

The geologic story of Bernheim Forest began more than 400 million years ago when the area was covered by great bodies of water or seas (Fig. 2). In the Bernheim area this lasted more than 100 million years. During this period various muds, sands, shell fragments, and lime oozes accumulated on the ocean bottoms much as they do today. Mud became clay and shale. Loose sand and silt became sandstone and siltstone. Shells, shell fragments, lime oozes, and chemical lime precipitates became limestone.

Rocks in Bernheim Forest are layered, like a cake. The lowest layer is the oldest (Ordovician) whereas younger rocks (Mississippian) cap the hills and ridges (Fig. 3). Ordovician limestones are found only in a few spots in remote areas on the eastern edge of the Forest; Silurian limestones and dolomites are the oldest rocks visible in the areas most frequently visited.

The variety of rock types found in the Forest suggests that the sediments were deposited under different conditions or environments. The limestones in the valleys and near the base of the hills were formed when warm, shallow sea water covered the region. The ocean bottom was inhabited by myriads of sea clams, corals, snails, and other creatures. Some of the shell and skeletal remains are preserved in the limestone rocks for us to see today (Fig. 4).

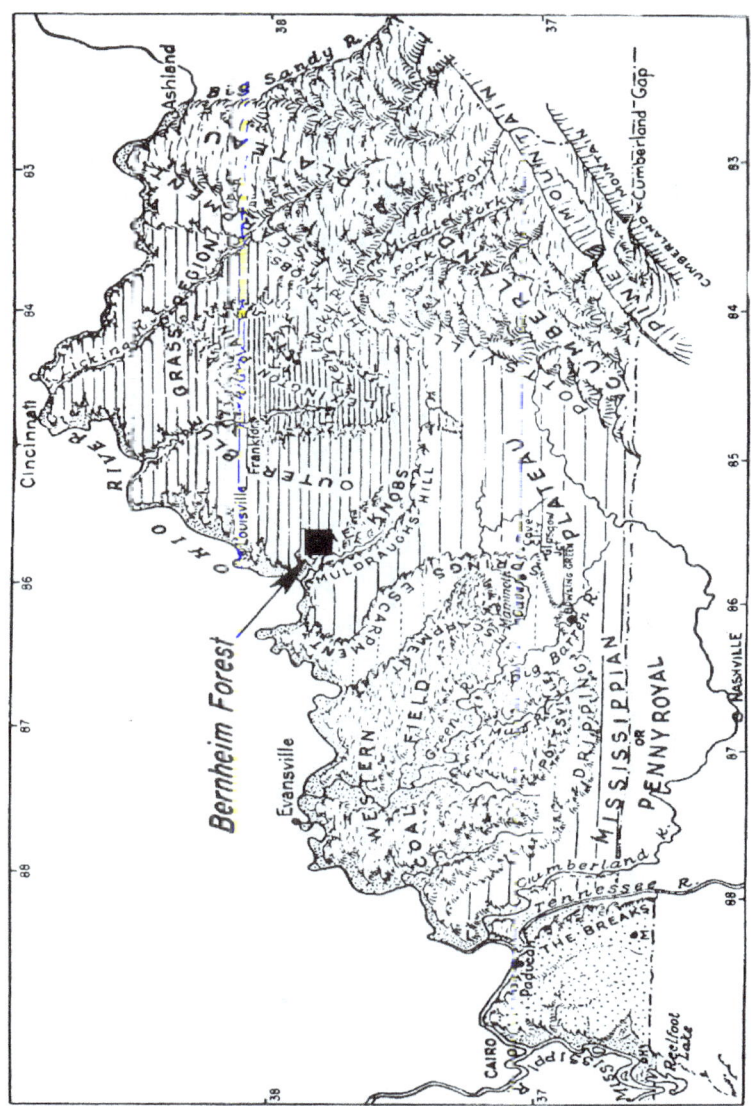

*Figure 1. Physiographic map of Kentucky (adapted from A. K. Lobeck, University of Columbia Press) showing the major geographic and geologic areas of the State and the location of Bernheim Forest.*

5

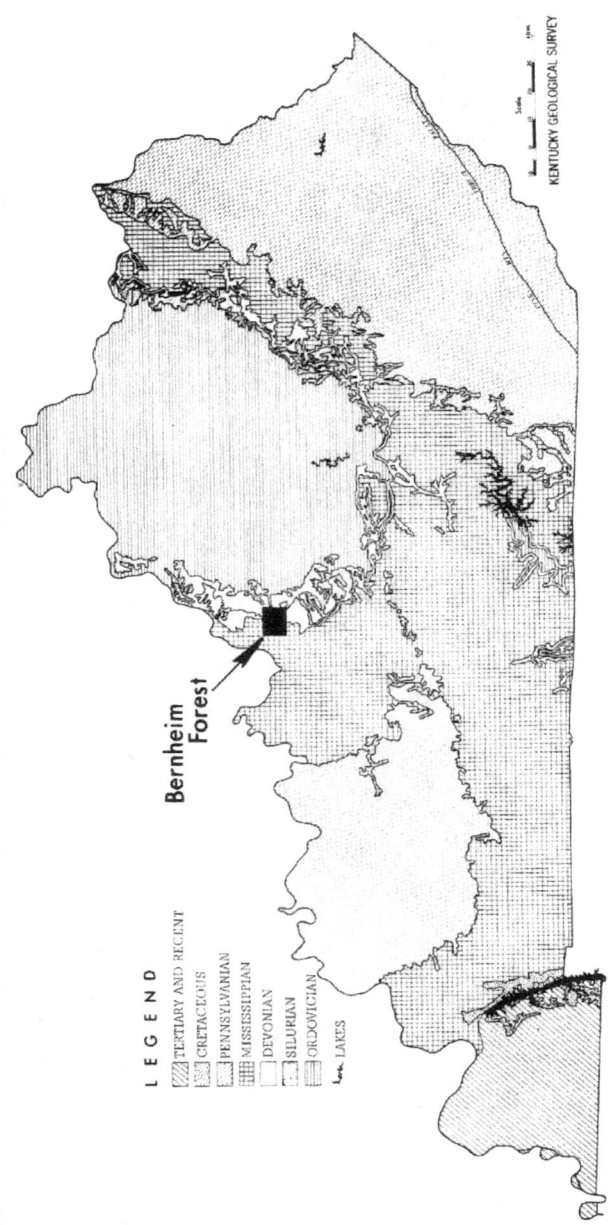

*Figure 2. Outline geologic map of Kentucky showing the distribution of rocks of various ages and the location of Bernheim Forest. The Ordovician rocks, which are exposed in*

6

*the Blue Grass regions, are the oldest. They dip gently to the west from Lexington; successively younger rocks are encountered on the surface of the ground as one proceeds in that direction. Thus, at Bernheim Forest one would have to drill a hole several hundred feet deep to reach the same rock formations seen on the surface at Lexington. Rocks in Bernheim Forest include Silurian, Devonian, and Mississippian ages.*

| ERA (DURATION IN MILLIONS OF YEARS) | ROCKS EXPOSED IN KENTUCKY | BERNHEIM FOREST | PERIOD (DURATION IN MILLIONS OF YEARS) | | AGE (MILLIONS OF YEARS) |
|---|---|---|---|---|---|
| CENOZOIC 64± | | | QUATERNARY | 1 | 1 |
| | | | TERTIARY | 62 | 63 |
| MESOZOIC 167± | | | CRETACEOUS | 72 | 135 |
| | | | JURASSIC | 46 | 181 |
| | | | TRIASSIC | 49 | 230 |
| | | | PERMIAN | 50 | 280 |
| | | | PENNSYLVANIAN | 30 | 310 |
| | | | MISSISSIPPIAN | 35 | 345 |
| PALEOZOIC 370± | | | DEVONIAN | 60 | 405 |
| | | | SILURIAN | 20 | 425 |
| | | | ORDOVICIAN | 75 | 500 |
| | | | CAMBRIAN | 100 | 600 |
| PRE - CAMBRIAN 4,000± | | | | | 4,600 |

*Figure 3. Geologic time chart showing the relation of the age of the rocks in Bernheim Forest to those in the rest of Kentucky. The scenic features of this interesting region have been formed in rocks which are 325 to 425 million years old.*

Later the sea floor became covered with an organic, black muck. This muck is now a hard black shale which geologists call "New Albany Shale" for the excellent exposures along the Ohio River near that Indiana city. In the vicinity of Bernheim Forest, it can be seen in numerous road cuts and also in the bed of Slate Run in the Forest itself (Fig. 5). Fossil remains of the earliest known trees are found in this formation (Fig. 6).

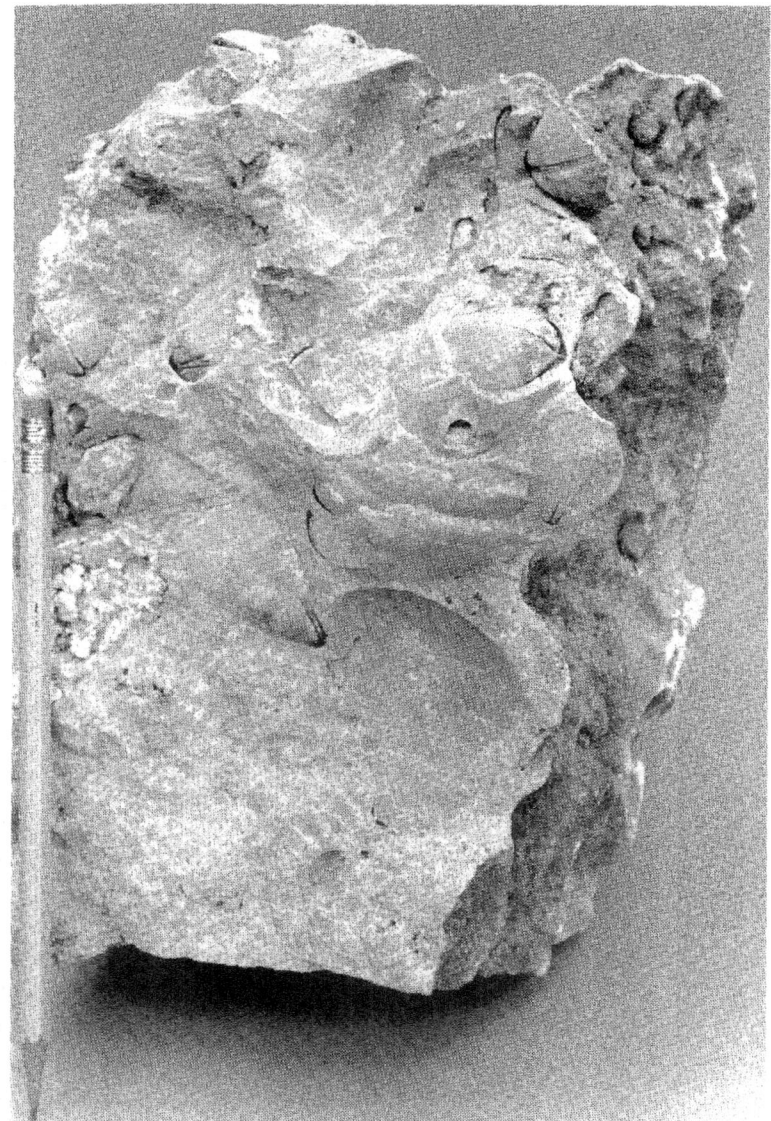

Figure 4. Block of Middle Silurian dolomite with casts and molds of the brachiopod Pentamerus. This is a guide or index fossil for these rocks.

Layers of shale and siltstone over the black shale formation tell us that the sea in which these sediments were deposited was muddy (Fig. 7). Some geologists consider these rocks to have been part of a great delta, formed by sediment carried by ancient rivers and streams from uplands many miles to the northeast and deposited in the Mississippian Sea. Peculiar markings on some of the slabs of siltstones are indications of water currents and sea bottom life.

These are not "just rocks" because several of them have special significance. For example, the Laurel Dolomite (Silurian), which was formed during the same geologic period as many of the ancient coral reefs, possesses characteristics which make it an excellent reservoir rock for some of nature's fluids. In the Bardstown area many wells which tap this geologic formation yield ground water for domestic and livestock purposes, whereas in Green County, Kentucky, where it is much deeper and in a different geologic situation, the Laurel Dolomite was the "pay zone" for the prolific Greensburg oil pool which yielded more than 18 million barrels of crude oil between 1958 and 1962. The stone is also quarried for use in road construction. A conspicuous spring zone occurs near the base of this formation throughout the outcrop area. The presence of these springs was one of the reasons the distilling industry was originally attracted to this part of Kentucky.

*Fig. 5. Outcrop of New Albany Shale near the mouth of Slate Run. Easily recognized by the black color and hard, brittle layers.*

The black New Albany Shale, where deeply buried, has produced natural gas in several areas in Kentucky and southern Indiana. Experimental work has indicated that this same shale when "cooked" yields 10 to 20 gallons of oil per ton of shale. While this is not a sufficient quantity of oil to be considered commercial at the present time, it may represent a reserve of an important mineral fuel for sometime in the future.

*Figure 6. A portion of a fossil tree found in Bernheim Forest. It has been identified as Callixylon newberryi, a form which is considered as probably the oldest fossil tree known, approximately 350 million years old. The geological processes which preserved this tree were probably similar to those which preserved the extensive ancient forests of our western states in the vicinity of Petrified Forest National Park. The tree fell or was washed into a bay where it was rapidly buried with mud and silt. The deposits in which the tree was buried eventually turned into shale which is now called the New Albany Shale. These sediments contained a large amount of silica. This silica was picked up by ground water, carried into the wood and deposited in the cell tissue. Eventually the mineral filled the wood almost solidly, forming the present petrified lag.*

The soft gray and green shales overlying the New Albany Shale are used in several localities for the manufacture of structural clay products. A plant in northern Bullitt County subjects the raw shale to rapid heating to produce a lightweight aggregate for use in concrete blocks and other construction purposes and as a mulch. A plant in Jefferson County uses a

11

similar deposit of shale for the manufacture of bricks used for facing buildings and homes.

*Figure 7. Outcrop of soft, gray and green Lower Mississippian shale. Shales similar to this are used in Kentucky for the manufacture of brick, lightweight aggregate, and other ceramic products. Sometimes small, round concretions containing phosphate are found near the base of this formation.*

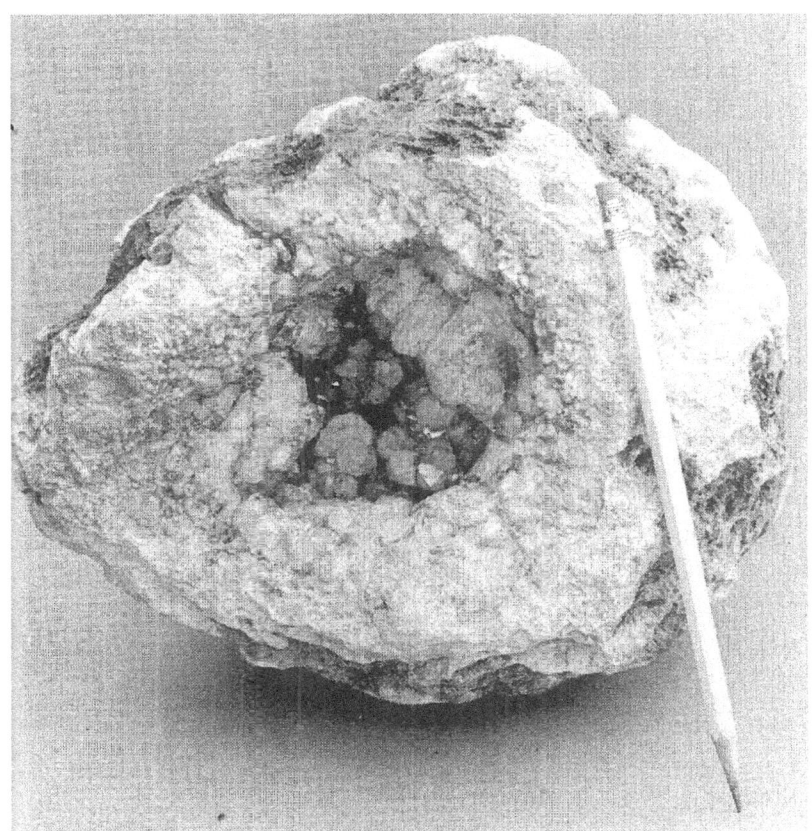

*Figure 8. Geode lined with chalcedony and quartz crystals. Geodes are commonly found in limestones of Mississippian age or accumulated in stream beds in and near the Knobs areas.*

Those hard, globular bodies of silica called geodes are of interest to many rock and mineral collectors, amateur and professional alike (Fig. 8). They are commonly found embedded in Mississippian limestone ledges or accumulated in stream beds in and near the Knobs areas. Usually well formed clear quartz crystals line the inside of the hollow geodes, but they may contain a variety of minerals. More than a dozen different minerals have been reported from Kentucky geodes; geodes filled with gypsum are common in adjacent Hardin County.

Of historic interest are the old iron ore "diggings" on the hillside northeast of the mouth of Wildcat Hollow. The ore obtained from this small mining operation occurs in the lower Mississippian shales as siderite (an iron carbonate mineral) which weathers out as limonite (Fig. 9). The siderite concretions were probably formed by water which soaked through the rock and deposited lime and iron compounds in concentric layers around a fossil or mineral fragment. The reddish brown concretions or nodules were exposed when streams cut through the layers of rock containing them. The ore from the locality was hauled to Bellemont (Belmont) furnace which was situated about a mile east of the present community of Belmont, southwest of Bernheim Forest. Limonite is a very low grade ore. For this reason and because it is present only in small tonnages in the area, it could not support much of an iron industry. Only an inconspicuous scar on a hillside indicates that mining took place in Bernheim Forest approximately a century ago.

*Figure 9. Limonite and siderite concretions in Early Mississippian shale. These were built up by water which soaked through the rock and deposited*

*iron and lime compounds in concentric layers around a fossil or mineral fragment. Deposits such as these provided the ore for the small iron furnaces which operated almost a century ago near Belmont and Nelsonville.*

## The Knobs

After the sediments which were deposited in the ancient seas became hard, the crust of the Earth was warped and raised slowly upward until the seas retreated permanently from Kentucky. One of the folds resulting from this warping or uplift was the Cincinnati arch which extends from Ohio southwest across Kentucky through Lexington and then toward Nashville. From the center of this great fold or arching, rocks dip gently essentially at right angles to the axis. Thus rocks found near the tops of the hills north of Bardstown will be in the valley bottoms in Bernheim Forest. And these rocks, in turn, will be several hundred feet deep near the entrance to Fort Knox north of Elizabethtown.

Subsequent periods of uplift during later geologic ages raised the land to approximately its present elevation above sea level. Streams then began their erosive work, carving valleys and sculpturing the landscape. It is continuing today.

In its typical development, the Knobs region is a narrow belt of country surrounding the Blue Grass, characterized by partially or completely isolated monadnocks in the form of low rounded hills or conical knobs capped by resistant layers of rock. These knobs often stand out as prominent landmarks (Fig. 10). Behind the Knobs in the Bernheim area is an eastward facing cuesta or escarpment capped with resistant limestone. Here it is referred to as Muldraugh's Hill; in nearby Indiana its counterpart is called the Knobstone Escarpment.

In the belt bordering the Cincinnati arch, hard resistant limestones and sandstones occur. These hard ledges of rock serve as a caprock to retard surface erosion and protect the softer shales and siltstones below. The Knobs are erosion remnants which have been detached from the main upland by

15

stream erosion. When they first become cut off or isolated they are flat topped hills or ridges. Further erosion removes the resistant cap and they take on a conical form (Fig. 11). These cone shaped peaks constitute some of the most striking topographic features in the Bernheim region.

## Topography

Bernheim Forest is a rugged upland area fairly uniformly and deeply dissected by stream action in a thick series of rock formations, resulting in a mature topography. The tops of the highest hills are more than 900 feet above sea level and the deeper valleys have been cut a few hundred feet below (Fig. 12).

The narrow, sharp ridges are the reverse of the deep stream trenches. In the upper portions, the valleys are V shaped but become flat bottomed where they join larger streams. Lateral or side cutting of the valleys has been rapid when the streams reached the shales and soft siltstones. No waterfalls of any consequence exist. The rather uniform height of the hills and ridges between the major streams presents an even skyline when viewed from the upland surface.

Salt River and Rolling Fork are the principal streams in this part of Kentucky. Waters falling in the Forest are drained by tributaries of these streams. Although the topography is rough, the principal streams occupy broad, flat valleys more than 400 feet below the upland. Salt River at Shepherdsville is 406 feet above sea level; tributary valley flats are between 425 and 500 feet above sea level.

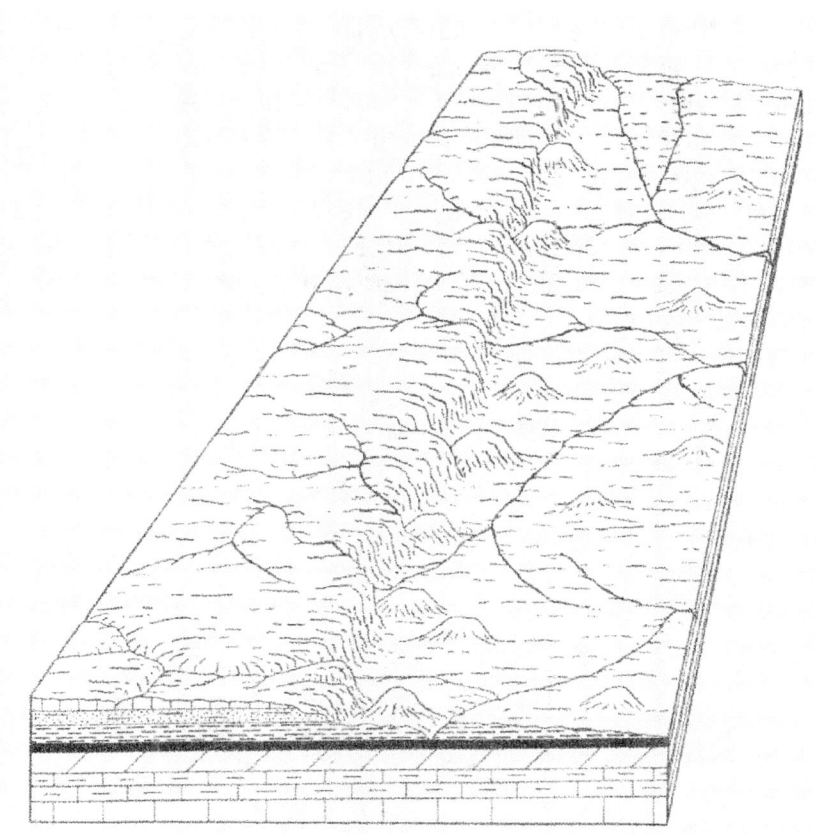

*Figure 10. Block diagram showing Muldraugh's Hill and development of the Knobs (adapted from McFarlan, 1958, p. 37). Muldraugh's Hill in this area is an east-facing cuesta or escarpment capped with resistant St. Louis and older Mississippian limestones. The Knobs are erosion remnants of the upland area after the front of the escarpment has been carved by stream erosion. When first cut away from the escarpment, the hill areas may be flat topped, but as the resistant cap is removed and softer shales and siltstones are exposed to weathering processes the hills take the shape of cones.*

Most of the ravines and valleys within the Forest proper are steep sided and V shaped in cross section. Some of the interstream ridges are flat or nearly so, representing remnants of hard layers of sandstone which resisted erosion. From these ridges minor spurs extend approximately at right angles to the streams. The ends of the spurs are not as high as the main ridges, but are commonly 150 feet or more above the valleys. Steep slopes are to be found everywhere, but rarely if ever an unscalable bluff.

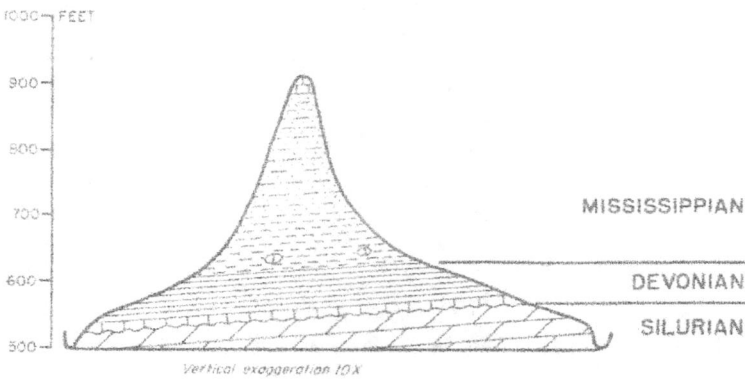

*Figure 11. Cross section of a representative knob. The resistant cap has been removed by erosion but hard siltstones in the upper part have prevented it from being destroyed entirely. The gentle slopes have been developed on the softer shales and silty shales.*

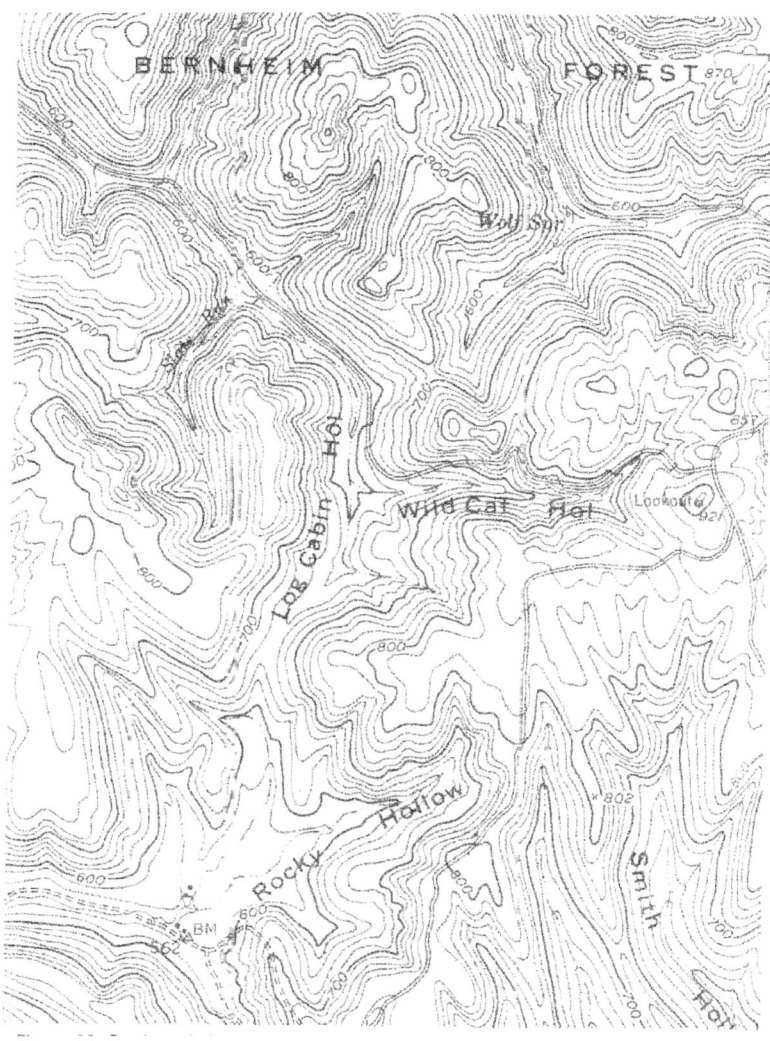

*Figure 12. Portion of the Shepherdsville topographic map showing some of the rugged topography in Bernheim Forest. More than 300 feet difference in elevation exists between the valley bottoms and the tops of the higher knobs. The knob near the head of Wild Cat (Wildcat) Hollow on which the lookout or fire tower is situated is 921 feet above sea level; this is one of the highest points in the Forest. Contour interval of the map is 20 feet.*

19

The rocks are sufficiently resistant to erosion to form the hills, ridges, and conical knobs, but they are also soft enough to weather easily and support forest growth. Thus their contrasting characteristics combine to produce a picturesque and scenic segment of Kentucky's landscape.

A profile of the geology and topography of the Forest can be seen from the entrance to the fire or lookout tower.

The entrance station is about 500 feet above sea level. Geologically, it is near the boundary between the Devonian limestones and the black shales, although soil and vegetation cover obscures the actual contact at this point.

Large blocks of limestone along the edge of Lake Nevin, near the Arboretum Center and along some of the trails, are of both Devonian and Silurian ages (Fig. 13). They have been placed there as a part of the landscaping activities. In this area, the limestones of Devonian age can be identified by the abundance of crinoid stem fragments preserved in the rock (Fig. 14).

*Figure 13. View of Lake Nevin, a man made impoundment in Bernheim Forest. The large rocks along the water's edge are limestones of Devonian and Silurian ages, placed here as a part of the landscape activity of the Forest. The wooded hills in the background are carved from Early Mississippian shales and siltstones.*

*Figure 14. Limestones of Devonian age with fragments of crinoid stems, parts of an ancient sea animal, preserved in the rock. These may be seen in the fountain at the Arboretum Center, in some of the boulders along Lake Nevin, and in the bed of Rock Run near the mouth of Slate Run.*

The large, nearly flat grassland area, called Big Meadow, is developed on the New Albany (Devonian) black shale formation as is the flat area on which the museum is located.

The numerous outcroppings of rock in the vicinity of the waterfowl area and along Cedar Lakes in front of the museum are Silurian limestones and dolomites.

From the museum to the lookout or fire tower several rock types may be observed in the streams and road cuts. Limestones are the predominant type in the lower part of Rock Run (Fig. 15). Some of the better exposures of the black New Albany Shale are found in Slate Run, a tributary of Rock Run. The name of this small stream undoubtedly came from the outcrops of the black, blocky and platy shale (Fig. 16). This is not a true slate, geologically speaking, but it is understandable how its physical characteristics have caused this mistaken identity.

As one begins the steep climb up the road, fragments of siltstone and silty shale begin to appear. These are from formations which lie above the black shales. The old iron ore pit, which provided ore for Belmont furnace in the 1800's, is found near here. Sandstones are exposed in the sharp horseshoe curve in the road. Elsewhere in the area this sandstone unit may be present at hilltop positions. It is so resistant to erosion that flat ridges occur which have not yet yielded to the erosion agencies that have sculptured the characteristic rounded hills of the region.

*Figure 15. Limestone ledges of Silurian age along Rock Run in Bernheim Forest.*

At the lookout tower, elevation 921 feet at the base, the visitor is rewarded with a panoramic view of the region. Beyond the forested hills, conical knobs rise above the general level of the landscape. And on a clear day, Muldraugh's Hill, one of the most prominent landscape features in Kentucky, is plainly visible on the western skyline.

*Figure 16. Outcrop of New Albany Shale in the bed of Slate Run in Bernheim Forest. The vertical fractures or joints form rectangular patterns and give the shale a blocky appearance.*

Hopefully, the visitor leaves the area with a greater appreciation of the role which rocks play in forming the landscape. Although possibly not as highly publicized as forest, water, and wildlife resources, Kentucky's geologic resources are equally important. Bernheim Forest constitutes a memorable facet of this aspect of the Kentucky scene.

# Appendix

Browne, R. G., 1958, The Geology of Bernheim Forest: The Kentucky Naturalist, vol. XII, no. 2, p. 27-53, plates.

Kepferle, R. C., 196_, Geologic map of the Samuels quadrangle: U.S. Survey Geol. Survey Geol. Quad. map, in preparation.

Kepferle, R. C., 196_, Geologic map of the Shepherdsville quadrangle: U.S.Geol. Survey Geol. Quad. Map, in preparation.

McFarlan, A. C., 1943, Geology of Kentucky: Lexington, Univ. Kentucky, 531 p.

McFarlan, A. C., 1958, Behind the Scenery in Kentucky: Kentucky Geol. Survey, ser. 9, Special Pub. 10, 144 p.

Miller, A. M., 1919, The geology of Kentucky: Kentucky Dept. Geol. and Forestry, ser. 5, Bull. 2, 392 p.

Peterson, W. L., 196_, Geologic map of the Cravens quadrangle: U. S Geol. Survey Geol. Quad. Map, in preparation.

Peterson, W. L., 196_, Geologic map of the Lebanon Junction quadrangle, U. S. Geol. Survey Geol. Quad. Map GQ-603, in preparation.

U. S. Geological Survey, 1962, Topographic map, Cravens quadrangle, Kentucky: scale 1:24,000.

U. S. Geological Survey, 1962, Topographic map, Lebanon Junction quadrangle, Kentucky: scale 1:24,000.

U. S. Geological Survey, 1962, Topographic map, Samuels quadrangle, Kentucky: scale 1:24,000.

U. S. Geological Survey, 1962, Topographic map, Shepherdsville quadrangle, Kentucky: scale 1:24,000.

## KENTUCKY GEOLOGICAL SURVEY

27

# LETTER OF TRANSMITTAL

April 10, 1967

Dr. Raymond C. Bard
Assistant Vice President for Research
University of Kentucky

Dear Dr. Bard:

Bernheim Forest is a nature area preserved for public education and enjoyment as an instructive outdoor museum which is visited by thousands of Kentucky school children and others each year. An understanding of the geology is desirable for the visitor to fully appreciate all the natural aspects of the region.

This is another of our popularized publications on the outstanding scenic features of Kentucky.

Respectfully,
WALLACE W. HAGAN
Director and State Geologist
Kentucky Geological Survey